MARTIN GARDNER'S

SCIENCE
Magic

TRICKS & PUZZLES

MARTIN GARDNER

WITH ILLUSTRATIONS BY
TOM JORGENSON

DOVER PUBLICATIONS, INC.
MINEOLA, NEW YORK

Bibliographical Note

Martin Gardner's Science Magic: Tricks & Puzzles, first published by Dover Publications in 2011, is an unabridged republication of the work originally published as *Science Magic: Martin Gardner's Tricks and Puzzles* by Sterling Publishing Co., Inc., New York, in 1997.

Library of Congress Cataloging-in-Publication Data

Gardner, Martin, 1914–2010.
    Science magic : Martin Gardner's tricks & puzzles / Martin Gardner ; illustrated by Tom Jorgenson.
        p. cm.
    Originally published: New York : Sterling Pub., c1997.
    ISBN-13: 978-0-486-47657-5
    ISBN-10: 0-486-47657-X
    1. Science—Experiments—Juvenile literature. 2. Scientific recreations—Juvenile literature. I. Jorgenson, Tom ill. II. Title.

Q164.G25 2011
793.8—dc22

                                                                    2011004938

Manufactured in the United States by Courier Corporation
47657X05   2013
www.doverpublications.com

Preface

The book you are holding is, as its title indicates, a collection of entertaining science tricks, stunts, and puzzles. For the most part they are so little known, even to science teachers, that they are described here for the first time. Many were first published in *Physics Teacher*, to which I contribute a regular feature called "Physics Trick of the Month." Some I wrote anonymously, many decades ago, for *Children's Digest*. Others appear here for the first time.

The periodical *Magic*—it circulates among magicians—has been regularly reprinting my *Physics Teacher* contributions, where they are handsomely illustrated by Tom Jorgenson. His *Magic* pictures are reproduced here, along with many new drawings made especially for this volume.

Science tricks, their origins unknown, are passed around from magician to magician. Conjuring has long been my main hobby, which is how I came across many of the tricks in this book. I can claim credit for inventing only three: the Cartesian Matches (first published in *Match-ic*, a little book of match tricks that I wrote in 1935); The Rising Marble; and the Magic Bird, designed for *Children's Digest*.

The hand symbol next to a trick means you may need a helping hand from an adult if you are a young reader, because matches, sharp knives, boiling water, etc., can be harmful.

—M.G.

Metric Conversion

Inches	mm	cm	Inches	mm	cm	Inches	cm
⅛	3	0.3	6	152	15.2	23	58.4
¼	6	0.6	7	178	17.8	24	61.0
⅜	10	1.0	8	203	20.3	25	63.5
½	13	1.3	9		22.9	26	66.0
⅝	16	1.6	10		25.4	27	68.6
¾	19	1.9	11		27.9	28	71.1
⅞	22	2.2	12		30.5	29	73.7
1	25	2.5	13		33.0	30	76.2
1¼	32	3.2	14		35.6	31	78.7
1½	38	3.8	15		38.1	32	81.3
1¾	44	4.4	16		40.6	33	83.8
2	51	5.1	17		43.2	34	86.4
2½	64	6.4	18		45.7	35	88.9
3	76	7.6	19		48.3	36	91.4
3½	89	8.9	20		50.8	37	94.0
4	102	10.2	21		53.3	38	96.5
4½	114	11.4	22		55.9	39	99.1
5	127	12.7					

Contents

Water
- Cartesian Matches, 7   • The Gorilla Effect, 8
- Somersault Shell, 9   • Two 10-Cent Betchas, 10
- A Water Transfer, 11   • Water Level Riddle, 12
- Two Corking Good Challenges, 13   • The Dancing Triangle, 14
- The Marble and the Cork, 15   • How to Measure Volume, 16
- Three Jets, 17

Air
- Invisible Glue, 18   • Three for Bernoulli, 19
- A Blow for Bernoulli, 20   • The Mysterious Balloon, 20
- The Unbreakable Balloon, 22

Fire
- Candle Seesaw, 23   • Miniature Rocket, 24
- Light from the Wrong End, 25

Heat
- A Penny for Your Thoughts, 26   • Psychic Motor?, 27
- The Twisty Snake, 28

Gravity
- A Snap and a Drop, 29   • The Biased Penny, 30
- Balancing Silverware, 31   • Make a Magic Bird, 32
- Three Drop Tasks, 33

Motion and Inertia
- The Falling Keys, 34   • Roly-Poly Folder, 35
- Stabbing an Eggshell, 36   • The Waltzing Eggshell, 37
- Puzzling Quarters, 38   • Bottle, Hoop, and Dime, 39
- Rotating Egg, 40   • The Rising Marble, 41
- Pool Hustler Scam, 42   • Which Thread?, 43
- Transporting an Olive, 44   • The Frustrating Papers, 45
- Swinging Cups, 45

Friction
- Balancing a Book, 48   • Climbing Bear, 49
- Curious Feedbacks, 50

Magnetism

• The Levitated Paper Clip, 51   • Floating Paper Cup, 52
• Psychokinesis?, 53   • Pill Bottle and Paper Clip, 54

Electricity

• Sneaky Switches, 55   • Electric Pickle, 56
• The Human Battery, 57

Sound

• A Puzzling Moo Horn, 58   • Music from Paper, 59
• Mysterious Spirit Raps, 60   • A Talking Machine, 61
• The Ghostly Glass, 62

Light

• Colored Shadows, 63   • A Mirror Paradox, 64
• The Pulfrich Illusion, 65   • Retinal Retention, 66
• A Blacklight Code, 67   • A QM Paradox, 68
• Another Mirror Paradox, 68   • Blacker Than Black, 69

Sensory Illusions

• A Square That Ain't There, 70   • Where Does the Water Go?, 71
• The Enchanted Die, 72   • Left- or Right-Eyed?, 73
• The Bent Playing Card, 74   • Zombie Glass, 75
• An Illusion of Weight, 76   • Magnetized Pencils?, 77
• The Puzzling Snap, 78   • Touching Hands, 79
• Funny Brush-Off, 80   • Sloping Teeth, 81
• Multiplying Marbles, 82

Probability

• Whirling Wire, 83   • The Bunch Effect, 84
• A Probability Swi   85   • Surprising Dice Bet, 86
• Are You Psychic?, 87

Dollar Bills

• Linking Paper Clips, 88   • See George Smile and Frown, 89
• Put George to Sleep, 90   • Turn George Upside Down, 91
• Support a Glass, 92   • The Balanced Half-Dollar, 93
• How Many Eyes?, 94   • Blow It Over, 95

Index, 96

Water

Cartesian Matches

An ancient toy called the Cartesian diver is a small hollow glass figure that moves up and down in a cylinder filled with water when pressure is applied or released to the air above the water. Here's an amusing way to demonstrate the same effect with two paper matches.

You'll need a plastic bottle with a tight-fitting cap. Fill the bottle to the brim with water; insert two paper matches, with heads down; and cap the bottle. If you now squeeze the sides of the bottle the pressure will force water into the fibers of the paper matches, causing them to sink. When you release the pressure, the matches rise. One match usually travels up and down ahead of the other. By adjusting the pressure, you can make each match move up and down as you please.

Magicians like to use matches with differently colored heads. Call one Mike and the other Ike. You can now make the matches respond to such commands as "Come up, Ike" and "Go down, Mike."

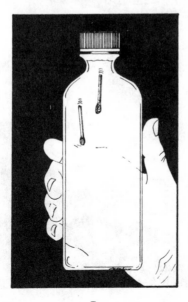

The Gorilla Effect

B efore showing this trick, secretly rub your wet *index* fingertip on a bar of soap.

Fill a shallow dish, a saucer will do, with water and scatter black pepper over the surface. The water represents a lake. The pepper grains are bathers. Your finger, you explain, models another bather entering the lake. So saying, touch the tip of your *middle* finger to the water at the saucer's rim. Nothing happens to the "bathers." Repeat this a few times to represent other bathers entering the water.

Now, you continue, along comes a gorilla who has escaped from a nearby zoo. Again, your finger models the gorilla as he enters the lake. This time, however, place the tip of your soaped *index* finger on the "lake's" edge. Instantly all the "bathers" flee to the opposite side!

Somersault Shell

C arefully open a fresh egg so that the two half-shells are as similar as possible. Check to make sure that the shell for the larger end has an air bubble inside. Most eggs do.

If the bubble is there, you can mystify a friend with the following stunt. Fill a tall glass with water. Give the shell without the air bubble to the friend, while you keep the other half-shell. Say nothing about the bubble.

Put your half-shell, open-side up, on top of the water, and gently push on the shell's rim until the shell fills with water and submerges. As it sinks, the bubble will cause it to flip over and land convex-end up. Fish it out with a spoon and challenge your victim to duplicate the feat. When he tries, his shell stubbornly refuses to turn over.

Repeat a few times. After the last somersault, surreptitiously poke your finger into the shell to break the bubble. If your friend thinks your shell differs in some way from his, let him now try it with your shell. To his puzzlement, the shell *still* refuses to flip over.

Two 10-Cent Betchas

A dime is on the table beside a glass of water. The glass must have straight sides. Hand someone a straw and say, "Betcha can't pick up the dime with this straw and drop it into the glass."

Here's how you do it. Put a drop of water on the dime. With one end of the straw in your mouth, bend over so the other end of the straw presses vertically on the dime. When you draw in air, the dime will adhere to the straw, allowing you to carry it over to the glass and let it fall in.

Follow this with, "Betcha can't drop the dime several inches to the table so it lands and stands on its edge."

Secret: Dip the dime in the water and push it against the outside of the glass near the brim. When you let go, the dime adheres to the glass, slides down the side to the table, and remains on its edge.

A Water Transfer

G lasses A and B are completely filled with water. B rests on two table knives, which in turn are placed across the brim of C. The bet is that you can transfer all the water from A to C without touching the glasses.

The secret? With a straw, blow vigorously at the spot where the brims of A and B touch. The air blast forces water out of A, between the brims of A and B. The water will drip down the outside of B and into C.

But how, I hear you asking, are you able to fill A and B? Simple. Just put the two glasses below the surface of water in a sink, press the brims together, and lift out!

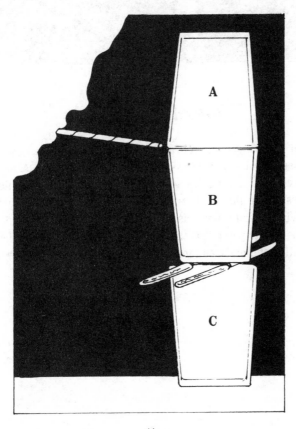

Water Level Riddle

F loat a small glass in a beaker filled with water, then add to the glass marbles, pebbles, or other small heavy objects, until the glass is close to sinking. Mark the water level on the beaker. Remove the glass, dump the marbles into the water, and refloat the empty glass. Will the water level rise or fall?

Few students will guess that the level will fall. It seems plausible that putting the marbles into the beaker would make the level rise.

Two Corking Good Challenges

Challenges

1. Pour water into a glass until it is almost full, and then drop a small cork on the surface. The cork will drift to one side. How can you make the cork stay at the center?

2. Fill a beaker or pan with water and float a cork on the surface. How can you make the cork sink to the bottom without touching it?

Solutions

1. Carefully add water to the glass until the surface becomes convex, rising slightly above the rim. The cork will move to the center or highest spot.

2. Hold an empty glass upside down over the cork and push it down. Air trapped inside the glass forces the water aside, allowing the cork to sink.

☞ The Dancing Triangle

Pour boiling water in a large bowl until it is about half an inch (1 cm) below the rim. Cover it with a cloth napkin or handkerchief. (Silk cloth works the best.) On top of the cloth put a triangle, cut from very thin paper, such as tracing paper, or paper used for second sheets in typing with carbon paper.

The paper triangle comes alive. It will curl up, then slowly uncurl and curl the other way, and keep up this curling and uncurling as long as the water remains hot. Its slow ballet dance may make it crawl over the edge of the bowl, but you can keep this from happening by constantly adjusting the cloth.

Try the experiment with a large shallow pan and a larger cloth. On top you can put papers cut into different shapes and watch them perform.

Why does this work? Moisture rising through the cloth expands the underside of the paper, causing it to curl; then the moist side exposed to the air dries out, the other side gets moist, and the paper curls the other way.

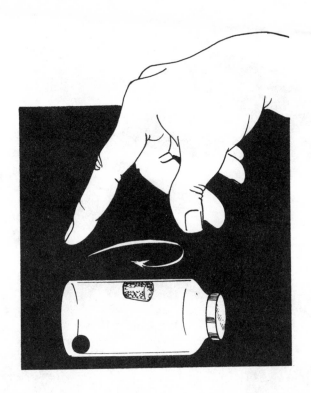

The Marble and the Cork

Put a marble and a cork (or a table-tennis ball) inside a jar completely filled with water. If you put the closed jar on its side and spin it, what will happen to the two objects? Curiously, the marble will roll to an end of the jar, as expected, but the cork moves to the jar's center.

The centrifugal force field in the spinning bottle is directed from the center (at zero radius) to the ends (at large radius). Objects that sink in fluids in gravitational fields will also "sink" in the centrifugal field. Hence the marble sinks toward one end or the other of the jar. The cork will float "up" toward the region of zero centrifugal field (and up in the gravitational field).

A similar effect is displayed by a floating balloon whose string is held by someone in the back seat of a moving car. If the car accelerates forward, the balloon moves forward, not back. When the car slows down, the balloon floats backward. And if the car rounds a curve, the balloon moves toward the curve's center.

How to Measure Volume

S how your audience two small figurines that are different, for example, a dragon and an elephant. Which has the larger volume? This may seem difficult to determine unless you think of this simple method. Plunge each into a beaker of water and see which figurine causes the water level to rise higher. Archimedes is said to have calculated the volume of a king's crown by this technique.

Three Jets

With a sharp pencil, punch three holes in the carton, as shown in the circle.

Keep three fingers against the holes; then fill the carton with water. Hold the carton over a sink and remove your fingers. The water will pour out in three streams, as shown on the left. The top stream will be weak, the middle one much stronger, and the bottom one the strongest.

Why it happens: The weight of the water causes a "pressure" in the water, which forces a stream out through each hole. The deeper the water, the greater the weight, and therefore the greater the pressure. The top stream comes from a shallow depth where the pressure is weak. The middle stream is stronger, and the bottom stream, of course, has the greatest pressure of all.

In lower depths of the ocean, the water pressure is so enormous that if deep-sea divers did not wear thick protective suits, it would kill them.

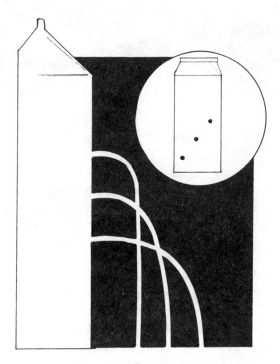

Invisible Glue

Arrange five playing cards face down in a row on a cloth surface. Moisten your thumb and press it firmly on the center of card A. When you raise your hand, A sticks to the thumb. Put A squarely on B, all edges flush, push down hard, and lift. B will adhere to A. Continue to the right until all five cards have been picked up in this way and placed on the palm of your other hand.

Nob Yoshigahara of Tokyo, who invented this stunt, likes to pretend he is putting a drop of invisible glue on the middle of each card before he does the pickups.

Air pressure keeps the cards together. The trick works only on a cloth surface or hard rug; otherwise the cards stick to the surface.

Three for Bernoulli

Bernoulli's principle states that air in rapid motion will lower the air pressure along its path. Here are three easy ways to demonstrate this.

1. Hold a drinking glass in one hand and with the other hand hold a burning match behind it. Blow toward the glass. The flame goes out as if you blew right through the glass.

2. Cut two long narrow strips from a newspaper. Bend your head forward and hold the ends of the strips on either side of your mouth. Blow down between the strips. You will see their lower ends come together instead of moving apart.

3. Push a thumbtack or pin through the center of a business card. Holding the card horizontally, place a spool on the card over the tack. As you blow down through the hole, let go of the card. You would expect the card to be blown off the spool, but instead it clings to the spool's underside.

A Blow for Bernoulli

There are many well-known ways to demonstrate the Bernoulli effect, but here's a simple demonstration that is not so well known.

At one end of a ruler, fasten a paper strip with a slight hump in the center as shown. The strip's ends can be pasted down or held with rubber bands. Place the ruler crosswise on a round-stemmed pencil. Roll the pencil back and forth until the end of the ruler with the paper strip very slightly overbalances the ruler's other end.

Blow toward the ruler. Air rushing over the strip's bulge will lower air pressure above the bulge and cause the ruler to tip the other way.

❖❖❖❖❖❖❖❖❖❖❖❖❖❖❖❖❖❖❖❖❖❖❖❖❖❖❖❖❖❖❖

The Mysterious Balloon

Can a toy balloon be inflated and remain this way even though its mouth stays open? Here's how to do it.

Put a few inches of water in a large glass bottle that has a small opening. Set the bottle on a low stove light. When the water boils, remove the bottle and very quickly attach the uninflated balloon's

mouth to the bottle's neck. The air inside the bottle contracts while it cools, drawing the balloon inside, turning it inside out, and inflating it. Now you have a curiosity to show friends. An inflated balloon with an open end!

If the balloon has printing on the outside, turn it inside out before you attach it to the bottle. The inflated balloon will then have its printing on the outside, and look even more puzzling.

There are two unusual variations of this trick that do not require heating. Obtain a large, hard-plastic bottle of the sort soft drinks now come in. Punch a hole near the bottom. You can now insert an uninflated balloon into the bottle, attach its mouth to the bottle's mouth, then inflate the balloon inside the bottle by blowing into it. The hole allows the air to escape as the balloon inflates. After the balloon is inflated, a thumb over the hole keeps the balloon from deflating. Exhibit this inflated balloon, with its open mouth, as a curiosity. Wave your hand over it, pronounce a magic word, and then take your thumb off the hole. The balloon instantly turns inside out to inflate again above the bottle.

A second variation, using the same hard-plastic bottle, is to attach the balloon to the mouth as before, with a straw alongside it. The straw allows air to leave the bottle as you blow to inflate the balloon. While the balloon is inflated, remove the straw, leaving the inflated balloon permanently inside the bottle.

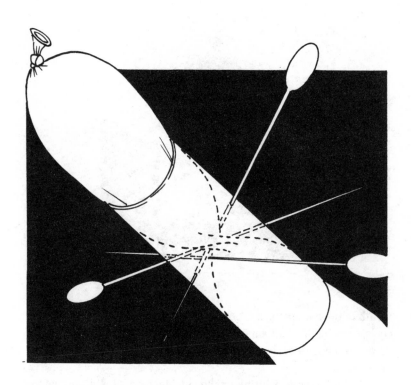

The Unbreakable Balloon

The elastic skin of a balloon is much thicker close to its neck, and also at the end opposite the neck than it is at other spots. A needle or a hatpin can be pushed into either of these spots without bursting the balloon. The holes are so small that the balloon seems to remain the same size after the needle is removed from the holes. Of course the balloon bursts if you stick the needle in at other spots.

An effective magic trick uses a balloon that inflates to a sausage form. Push it into a cardboard tube such as one that comes inside toilet paper rolls. As you push the balloon through the tube, secretly give it a twist. You now can push hatpins through the tube's middle as shown, taking care not to go directly through the balloon's twisted portion. For people who do not know that the balloon has been twisted, its failure to burst is mystifying. Withdraw the hatpins and secretly untwist the balloon as you remove it undamaged from the tube.

Fire

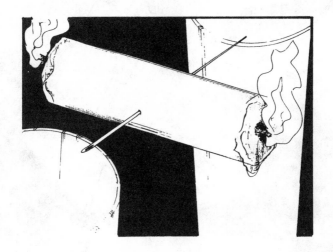

☞ Candle Seesaw

I burn my candle at both ends.
It gives a lovely light.
Up and down, up and down,
It seesaws through the night.

Cut away the tallow at the bottom end of a cylindrical candle to expose the wick. Push a needle through the candle's exact center. Rest the needle's ends on the rims of two glasses as shown, and put a plate under each end of the candle to catch wax drippings.

Light the candle at both ends. Give one end a slight push and the candle will start rocking up and down like a seesaw. The motion continues as long as the ends burn.

You might suppose that each time a candle end leaves a bit of wax on a plate it makes that end a trifle lighter, and that these changes in weight operate the little machine. But is this the case? Are the weight differences sufficient? Is there a way that the formation and dropping of the tallow can deliver a slight upward recoil? To find out you must burn your candle at both ends.

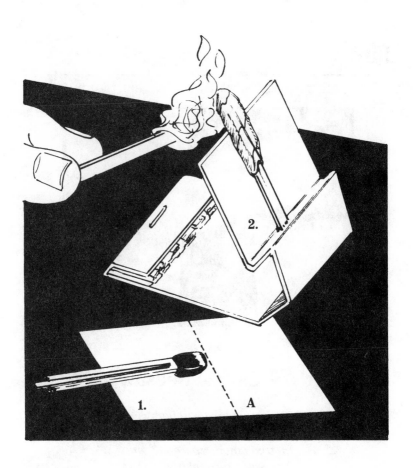

☞ Miniature Rocket

Place a toothpick alongside a paper match. Under the top two-thirds of the match put a piece of tinfoil about an inch by two inches, as shown in Figure 1. Fold the rectangular foil over the match, along dotted line A; then wrap the foil firmly around the match. Withdraw the toothpick. This leaves a tiny open channel running to the match's head.

Bend an *empty* match folder in the middle (Figure 2) so it can serve as a launching pad. Light another match and hold the flame beneath the foil. When the matchhead ignites, gas escaping through the channel will propel the wrapped match six or more feet across the room.

☞ Light from the Wrong End

Light a paper match and hold it vertically in your right hand. Hold an unlit paper match horizontally in your left hand as shown. You must be in a room where the air is totally free of drafts.

Bring the burning match over to the unlit one, and touch the flame to the head of the unlit match to light it. Now move your right hand away a foot or so, lean forward and blow out the match in your left hand. Bring the burning match back to the match you just extinguished, and touch the bottom of the burning match to the unlit head. It relights! Throughout these moves, your left hand must *not* move. Only the right hand moves away and then back.

Why does it work? There is a combustible component in the wisp of smoke that rises from the blown-out match. The flame jumps down this wisp so rapidly you can't see it. It looks exactly as if the bottom of the burning match relights the other one.

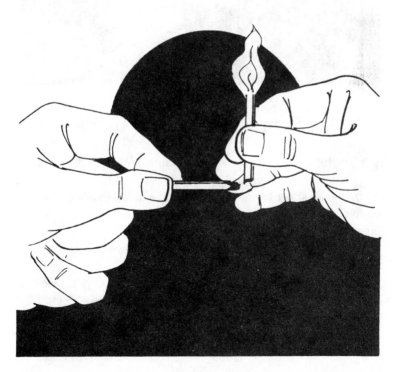

Heat

A Penny for Your Thoughts

Place ten pennies in a row on the table. While you go out of the room (or turn your back) someone is asked to select a penny, press it against her forehead with the tip of a forefinger, and hold it there while she counts slowly to fifty. The penny is then replaced in the row.

Explain that thought vibrations from her brain will be transmitted to the penny.

Return to the room. Touch a fingertip to each penny to detect its paranormal "vibrations." Sure enough, you correctly identify the penny she chose.

How do you do it? The selected penny will be distinctly warmer than the others.

Psychic Motor?

S tick a pin through a piece of cardboard. Fold a 2-inch square of paper in half both ways; then balance it on the pin's point as shown.

Hold your cupped hands on both sides of the tiny "motor." In your mind order the paper to rotate. Be patient. The chances are high that in a minute or two the little square will start turning. Try commanding it to stop and go the other way.

Psychokinesis? No. It is almost impossible for a room, even with windows closed, not to have convection currents arise from temperature variations in the air. The heat of your hands, even the heat of your body, can produce such currents. Breathing may also play a role.

It is hard to comprehend, but there are some fringe parapsychologists who actually believe that psychic motors of this sort (they can have many variations) are moved by paranormal forces!

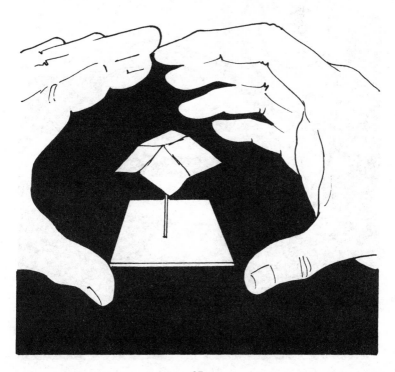

The Twisty Snake

D raw the pattern of the snake on a piece of cardboard (see circle). Cut along the black lines. Punch a hole at the spot in the center. Run one end of the piece of thread through the hole and secure it with some cellophane tape on the underside.

Hold the other end of the thread, so that the snake hangs in a spiral as shown. Hold it above a burning lamp. Soon it will begin to rotate slowly.

Why it happens: When air is heated, it expands and rises, making an upward convection current. Every burning light bulb has such a current above it, and your spiral snake will prove it! It is this upward draft of hot air that strikes the sloping underside of the snake and makes it turn.

Gravity

A Snap and a Drop

Balance a playing card on the tip of your left index finger and place a coin that is about 1 inch (2.5 cm) wide, such as a half-dollar or quarter, on top of the card. With the middle finger of your other hand, you can snap away the card, leaving the coin balanced on the fingertip. This is like pulling the tablecloth from under the dishes.

Glue or tape a quarter to the center of an index card. Attach another quarter to one short end of another card.

If you hold each card with its face parallel to the floor and drop them simultaneously, which card will hit the floor first? Because both cards weigh the same, should they not strike the floor at the same instant? No, the card with the coin on its end falls much faster.

Why? Because the coin's weight (downward) at one end of the card and the force of air (upward) at the other end turns the card vertical. It falls with almost no air resistance. The other card remains horizontal. Strong air resistance slows it down like a parachute.

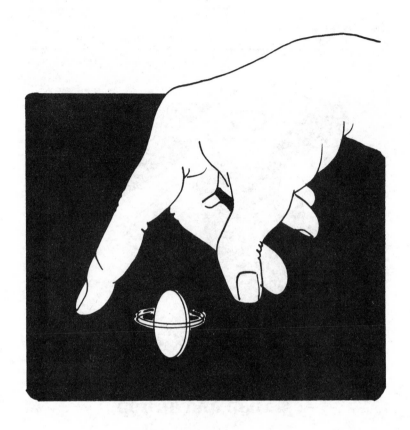

The Biased Penny

I 'll bet you didn't know that if you spin a shiny new U.S. penny on a glass or plastic tabletop, when it stops spinning it will fall heads up about 8 out of 10 times. I assume this is because when the Mint stamps pennies out of zinc sheets, before coating them with copper, an imperceptible bias occurs on the rims.

You can borrow anyone's penny, provided it is a new one, find a perfectly smooth surface to spin it on, and consistently win by betting on each spin that the coin will fall heads up.

A more striking way to demonstrate the tendency of new pennies to fall heads up is to balance 10 of them on their edge. Give the table a strong blow to make the coins topple. You'll find that a majority of the pennies, at times all 10, will fall heads side up.

Balancing Silverware

M esh together the prongs of two forks, then wedge a toothpick between them as shown on the left. You will discover that if you place the free end of the toothpick on the rim of the glass, the forks will balance as shown. Since the entire weight of the two forks is on the outer end of the toothpick, why doesn't it fall?

The answer is that the center of gravity of the entire "system" (forks and toothpick) actually lies directly beneath the point at which the toothpick rests on the glass. This is the lowest point that the center of gravity can assume, so it places the whole system in what physicists call a state of stable equilibrium.

The experiment illustrates the fact that a center of gravity may lie *outside* a physical object. The center of gravity of a rowboat, for example, is not a point within the wood, but a point somewhere in the space inside the boat.

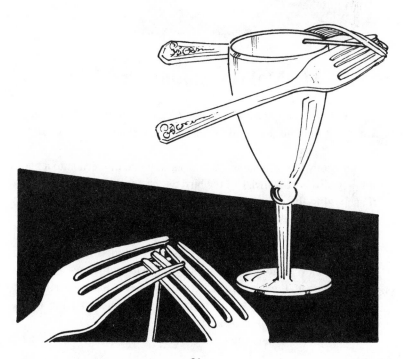

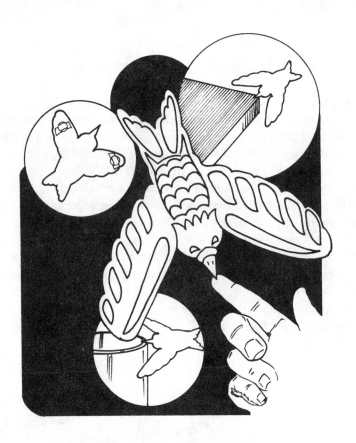

Make a Magic Bird

H ere's an easy-to-make cardboard bird that will balance on its beak in a magical and mystifying way. It seems to defy all the laws of gravity!

Photocopy the bird shown at 200% and paste the bird onto a piece of fairly stiff cardboard. Cut the bird out carefully. On the underside of each wing tip, glue a penny as shown in the left circle, or you can fasten the pennies with cellophane tape instead of glue.

Turn the bird right-side up again. Place the tip of its beak on the tip of your finger. The bird will remain suspended in a horizontal position! Show it to your friends. They will not know about the pennies beneath the wings, and will never guess what makes the bird remain balanced so mysteriously in mid-air. It can also balance on a glass rim or on a table's corner.

Make several of these magic birds and paint them in bright colors.

Three Drop Tasks

1. From a height of 6 inches (15 cm) above a table, drop a small matchbox so it lands on its end and remains upright.

2. Drop a cylindrical cork so it lands and stays on one end.

3. Drop a paper match so it lands and stays on its edge.

All three feats are impossible unless you know the method:

1. Secretly push the matchbox drawer about an inch (2.5 cm) upward as you hold the box vertically. The projection can be concealed by your hand. When the box hits the table, the inertia of its drawer sliding back in place will keep the box balanced on end.

2. Drop the cork on its side. After a few attempts, it will bounce up and land balanced on its end.

3. Bend the match in the middle before you drop it (see figure).

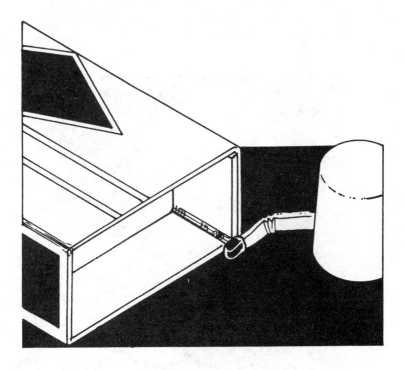

Motion and Inertia

The Falling Keys

A t one end of a 4-foot (3 m) length of string, tie two or three keys (or any small object of comparable weight). At the other end tie an empty match folder. Place the string over a pencil or pen (one with ridges, not a smooth one) so that the keys hang down about three inches from the pencil, while you hold the folder in the other end as shown.

If you let go of the folder, what will happen to the keys? Most people are sure the keys will fall to the floor. Incredibly, they don't! Try it and see what happens.

Most science tricks circulate from person to person and no one knows when or where they originated. This is a rare exception. It was invented by Stewart James, who first explained it in a 1926 magazine called *The Linking Ring*.

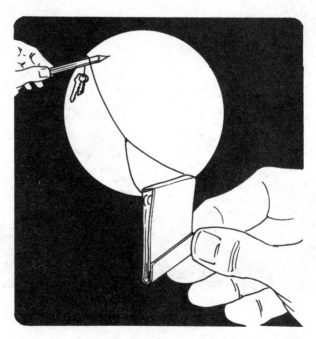

34

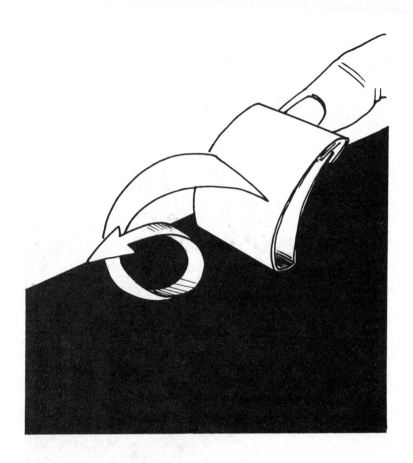

Roly-Poly Folder

T ear all the matches from a folder, then bend the folder into the curved shape indicated. Place it on a tablecloth and put the tip of a finger under the end with the striking surface. Raise the folder to a vertical position until it tips forward, free of the finger. Don't push. Just raise the folder until it overbalances.

The folder will turn a complete somersault, back to its original position. If someone wants to try it, you can play a trick on her by turning the folder end-for-end so her finger will be lifting the wrong end of the folder. Of course it won't somersault because the weight is at the wrong end. It's the inertia of the excess mass that causes the whimsical flip-flop.

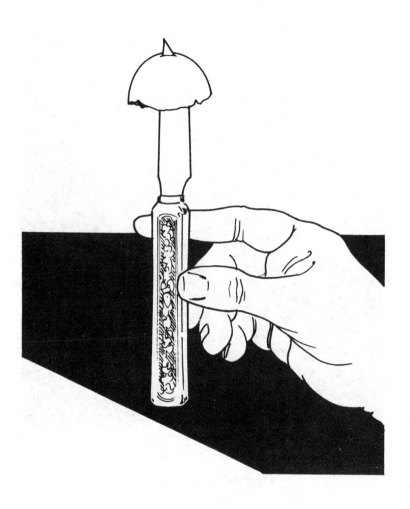

☞ Stabbing an Eggshell

Put half an eggshell over the end of a sharp-pointed kitchen knife. Problem: Tap the handle on a table or counter and cause the point of the knife to penetrate the shell. When others try it, the shell refuses to crack. Even after you show how it's done, your friends will still be unable to do it unless they are very observant. **Secret:** Appear to tap the handle on the table, but actually hold it loosely so that when the knife strikes the table it bounces.

The Waltzing Eggshell

From the side of an eggshell, break away a piece about the size of a half-dollar, about 1⅛ inches (3 cm) wide. Get a plate with a few inches of perfectly *flat* rim. Dip the plate in water; then place the broken shell on the edge of the rim, convex side down.

Tip the plate about 45° so that gravity starts the shell sliding down the rim. As it slides, it will start spinning rapidly. As it spins, keep rotating the plate clockwise in your hands. This keeps the shell spinning merrily while the plate turns beneath it. The moisture on the plate keeps the rotating shell attached to the rim.

You may have to experiment with more than one broken piece of shell until you find one that performs properly.

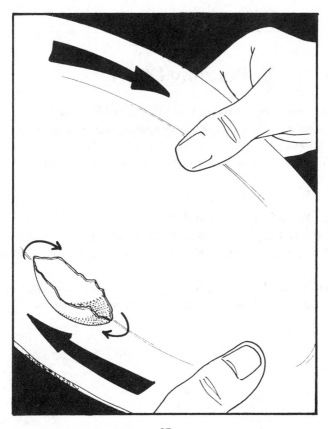

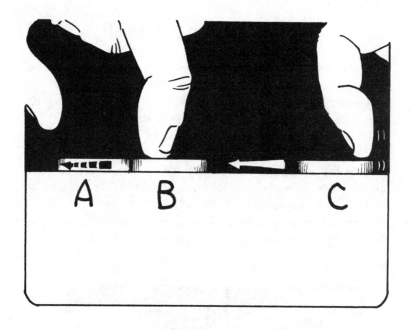

Puzzling Quarters

Put two quarters or other 1-inch (2.5 cm) wide coins side by side on a hard, smooth surface. Place a third quarter about 6 inches to the right as shown.

The task is this—put coin C between coins A and B without violating two rules:

1. You must not *touch* coin A.

2. You must not *move* coin B.

Solution: Put your left fingertip firmly on top of coin B; with your right fingertip flick C toward B. It is essential that C leave your finger before it strikes B.

When C hits B, its energy is transmitted to A, and A will slide to the left. This of course allows you to put C between A and B.

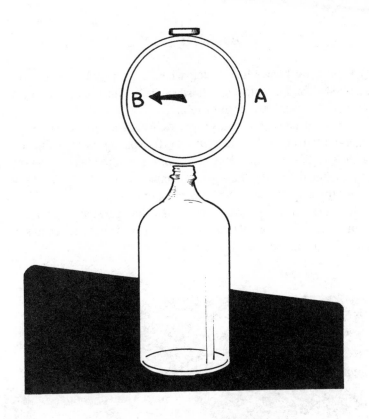

Bottle, Hoop, and Dime

Balance a wooden embroidery hoop on top of an empty bottle and place a dime or other small coin that is about ⅝ inch (1.5 cm) across on the hoop as shown.

The problem: knock the hoop away with a finger so that the dime falls into the bottle. When someone tries it, the hoop buckles into an oval that sends the dime flying across the room.

After everyone tries and gives up, you demonstrate how easily it is done. Instead of striking the hoop at point A, as your hand moves in the arrow's direction let your finger miss that side of the hoop and strike it *inside* at point B. The oval shape of the buckled hoop is now horizontal instead of vertical, allowing the dime to fall straight down into the bottle.

If you like, you can stand a large nail on the hoop instead of the dime.

Rotating Egg

Everybody knows how to tell whether an egg is raw or hard-boiled. Simply spin it on a hard surface. A hard-boiled egg spins merrily. Raw eggs are hard to spin.

Here's an egg stunt that is less well known. Give a raw egg a vigorous spin on its side. Instantly stop the spin by placing a fingertip on the egg, then just as quickly lift your finger. Although the pressure of your finger brought the egg to a complete stop, when you lift your finger the eggs starts slowly to turn again!

It is easy to understand why. When the egg stops rotating, its liquid interior continues to move with inertia strong enough to start the egg turning again.

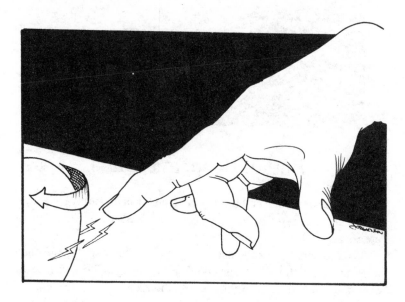

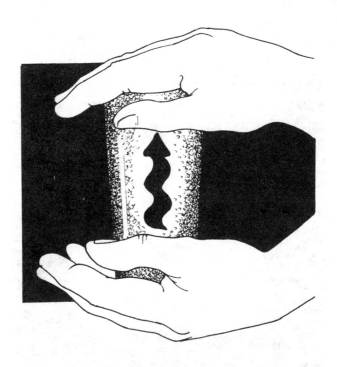

The Rising Marble

If you have a box filled with a mixture of large and small nails, you might think that shaking the box would tend to send the heavy nails to the bottom. Wrong! The reverse is true. The heavy nails rise to the top.

The phenomenon is easily demonstrated with dry rice and a large marble, glass bead, or any small, heavy object. Put the marble at the bottom of a drinking glass and pour in rice until the glass is full. Rest the glass on one palm; then tap the brim rapidly and vigorously with your other palm. In less than a minute you'll see the marble rise to the surface!

The trick works with almost any other type of small dry particles such as sand.

In 1995, a puzzle was on the market consisting of a glass tube, closed at both ends, filled with sand. The task was to move a steel ball through the sand from one end of the tube to the other end. The only way to do it is to hold the tube vertically, with the ball at the bottom, and shake the tub vigorously until the ball rises to the top.

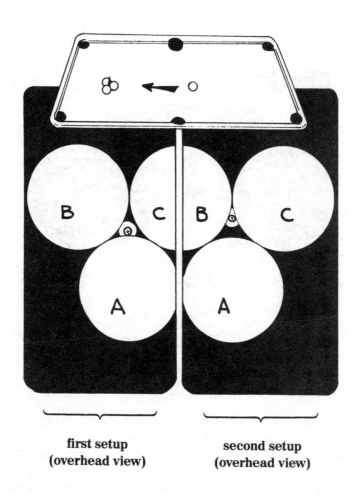

first setup
(overhead view)

second setup
(overhead view)

Pool Hustler Scam

P lace three pool balls at one end of the table as shown. In the tiny
space between them, stand a nail on its head. When you aim the
cue ball at the triangle of balls, it knocks over the nail. Set it up again
and challenge anyone to do the same. They find it impossible.

When you place the nail the first time, make sure that it touches
ball A. The cue ball will have enough momentum to knock over the
nail. When you set it up the second time, place the nail so that it
touches balls B and C but not ball A. Now when the cue ball hits A,
balls B and C will separate without disturbing the nail.

Which Thread?

Tie one end of a length of thread to a moderately heavy object such as a shoe or book, and suspend it in midair. Tie another length of thread to the object, letting it hang down as shown.

If you grab the lower end of the thread and give it a sharp yank, which thread will break? Because the weight of the object puts tension on the upper thread, one might expect it to break; however, the object's inertia, which keeps it from moving, causes the bottom thread to break.

The inertial principle involved here also explains the stage feat of placing a huge rock on a reclining person's chest, and then breaking it with a hammer without injuring the person.

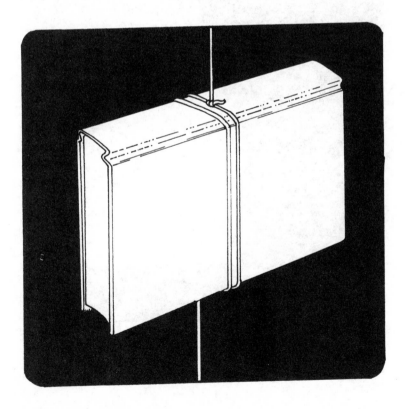

Transporting an Olive

G et two goblets with mouths a bit smaller than their middles. Stand one upright; invert the other over an olive.

Problem: Transfer the olive to the upright glass. You must touch nothing but the inverted goblet, and it must stay upside down at all times.

Secret: Rapidly rotate the inverted goblet in small circles. The olive will rise into the goblet, and centrifugal force will keep it there. Keep the goblet rotating until you can drop the olive into the other glass.

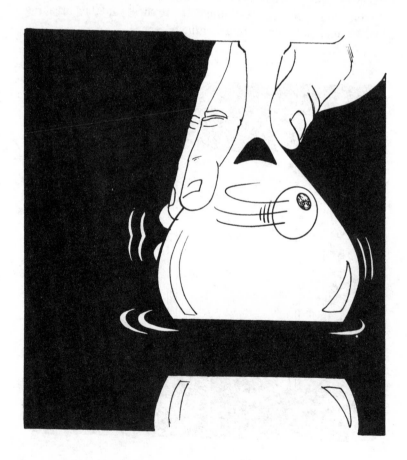

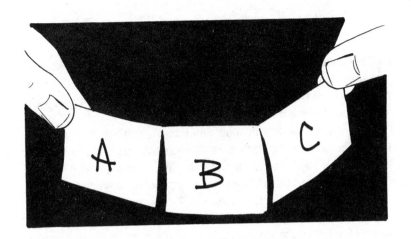

The Frustrating Papers

Tear a piece of paper, about 6 × 2 inches (15 × 5 cm) almost into three parts, as shown. They should hang together at the corners by the merest possible thread. Hold *A* in one hand, *C* in the other. Try to jerk them apart so that all three pieces separate. It's not possible! The paper will break at one spot, not at the other.

You can cheat by holding *B* with your lips, but there is a more interesting way. Using tape or a paper clip, attach a half-dollar to the top of *B*. This increases *B*'s inertia enough to cause the paper to break at both spots.

Swinging Cups

T his is surely one of the strangest magic tricks ever invented. The magician ties a cord between the backs of two chairs as shown. From this cord two coffee cups hang suspended on shorter cords. He starts the cup on his right swinging like a pendulum.

"Now watch closely," the magician tells his audience. "By concentrating the powers of my mind, I will cause this cup to stop swinging and the other cup to start." He waves his fingers mysteriously toward the cups. Sure enough, the cup on the right slows down until it stops completely. Meanwhile, the other cup begins swinging vigorously!

A moment later, the magician waves his hands again. This time he causes the left cup to stop swinging and the right one to start again. In fact, he can keep changing the motion from one cup to the other as long as the cups continue to swing!

The setup: Place two chairs back to back about 2 feet (61 cm) apart, and tie a string from the top of one chair to the top of the other. By means of shorter pieces of string, suspend the two cups as shown above. The measurements shown in this picture are important. The string holding a cup should be about 14 inches (35.5 cm) from the horizontal cord to the cup's handle. The two strings should hang about 6 inches (15 cm) apart. Adjust the chairs so that the center of the horizontal cord sags about 4 inches (10 cm) below the spots where the cord is tied to the chair backs. Now you are all set to demonstrate the power of your mind over the material objects.

How it's done: The secret of this trick is that there is no secret! There is really nothing for you to do except to *act* as if you are doing something. The trick works automatically, all by itself.

If you start one cup swinging at right angles to the horizontal cord, it will gradually lose its energy to the other cup. After the other cup has been swinging for a while, it will transfer its energy back again to the former cup. This back and forth transfer of motion will keep up as long as the cups are swinging.

What you must do, of course, is watch the cups carefully. As soon as you see a swinging cup start to slow down, raise your hands and wiggle your fingers toward the cups like a hypnotist, pretending that you are influencing them by your thoughts. No one will suspect that

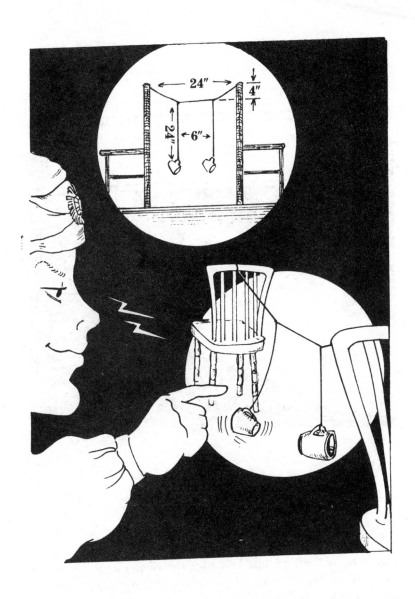

the trick would work just as well even if you were in another room!

You don't have to use cups for the trick. Any two small objects will work just as well, provided that they are not too light and that they weigh exactly the same amount.

Friction

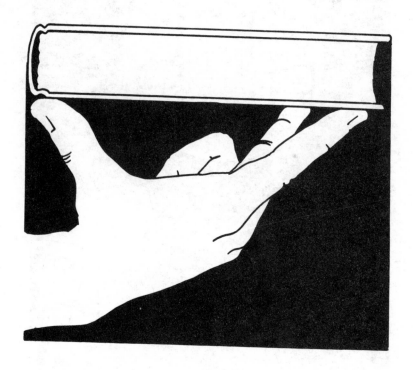

Balancing a Book

S pread your thumb and first two fingers as far apart as possible to form the points of a triangle. Place a hardcover book or metal tray on top of the fingertips. Regardless of where your fingers touch the book (provided its center is inside the triangle), when you bring the fingers together they will meet at the book's center, and the book will remain balanced. Try it with a large plate, holding it over a bed just in case.

Why does it work? Feedback, of the sort that operates flywheels and thermostats, does the trick. The closer a fingertip gets to the book's center, the heavier the weight on it and the greater the friction. This allows the fingers farther from the center to slide more easily.

Climbing Bear

U sing carbon paper, trace a bear on an index card, as shown, and punch four small holes at the black spots in the paws. Tie a loop in the middle of a length of string, and run the two ends through the holes (see the left drawing). Hang the loop on a projection or hook.

Lower the bear near to the ends of the string. Pull first on one string end and then on the other. The bear will waggle side to side and slowly climb to the top.

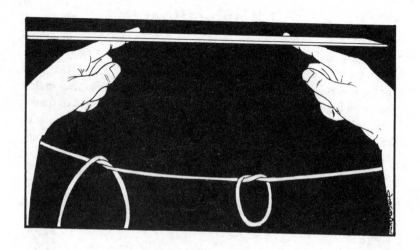

Curious Feedbacks

R est a yardstick horizontally on your extended forefingers. Move your hands slowly together. Your fingers will always meet exactly at the stick's center. Repeat with a small weight attached to one end of the stick. Your fingers will meet at the system's center of gravity, where the stick balances. Friction provides the feedback. If one finger gets ahead of the other, weight on it increases. This increases friction on that finger enough to keep it from moving, while the other finger, with less weight and friction, catches up.

Less well known is a similar stunt with a rope. Tie a large open knot near one end of the rope, and a more closed knot near the other end, as shown. Pull slowly on the two ends. The knots will close tight at the same instant. A longer rope, with knots of five or six sizes, makes a neat classroom demonstration as two students pull on the rope's opposite ends.

Magnetism

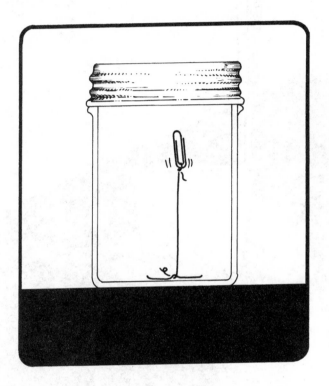

The Levitated Paper Clip

You need a glass jar with a metal cap that screws on. Tie a paper clip to one end of a short length of thread. Tape the other end to the bottom of the jar.

The thread should be of such length that if a magnet is stuck to the inside of the cap, the paper clip will be suspended in air like the Hindu rope trick.

When you show the jar to your audience, have the clip lying at the *bottom* of the jar. Let everyone examine the jar, but not the cap.

Turn the jar upside down so the clip hangs close to the cap. Then turn it right-side up again; the clip will remain suspended as you wave your hands over the jar and command it to levitate.

The Floating Paper Cup

S how your audience an opaque paper cup in your left hand. Under your left thumb, in back of the cup, is a small magnet.

With your right hand, lower a steel table knife into the cup. The magnet will cling to it, through the paper, allowing the cup to remain magically suspended in the air as you remove your left hand.

Finish by holding the cup with your left hand cupped over the bottom. As you remove the knife, the magnet drops into your left hand. Pocket the magnet as you hand the knife and cup out for examination.

Psychokinesis?

For more than a century, psychic charlatans have demonstrated their ability to influence a compass needle by waving a hand over it and commanding it to move. Here's how you can do it.

The secret is a strong magnet, concealed either in the tip of your shoe or strapped to your leg just above the knee. The compass is on a table where you sit. To influence the needle, raise your knee until it touches the table's underside near the compass, or lift your foot to bring the tip of your shoe under the compass.

Another spot for the magnet is under your shirt collar. As you keep ordering the needle to move, perhaps squeezing a fist over the compass to concentrate psychic energy, bring your face closer and closer to the compass until the hidden magnet under your collar affects it.

A subtle dodge is to sit at the table, keeping the secret magnet under the table at one side of the compass needle. This of course alters its direction. Onlookers note the direction of the needle. The compass is covered with a handkerchief. You then get up, walk to a distant spot, and pretend to use your psi powers to change the needle's direction permanently. The cloth is removed. Lo and behold, the needle is now pointing in a different direction—the direction in which it would normally point.

Pill Bottle and Paper Clip

D isplay a small pill bottle with transparent sides. It is filled with water, and inside is a metal paper clip.

The problem is to get the clip out of the bottle without spilling any water. You do it quickly behind your back.

A magnet is in your back hip pocket or under your belt. Remove the cap of the bottle and use the magnet to draw the clip up and out of the water. Replace the cap on the bottle and return the magnet to your hip pocket.

You can now show the bottle, still filled, and the clip in your hand. Onlookers will be mystified.

This trick can be done with several clips in the bottle, or with BBs instead of paper clips; it also can be done with a Canadian nickel, which is attracted to magnets.

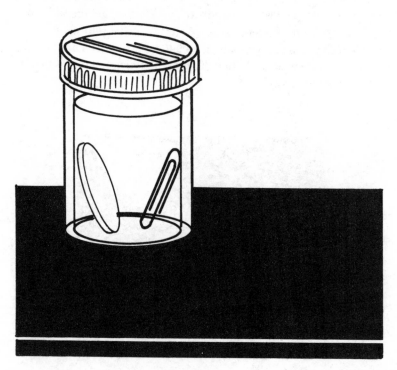

Electricity

☞ Sneaky Switches

Magic supply stores sell this bewildering trick, but it is easy to make your own apparatus. Mount a small light bulb on a board. Below it are three two-way push-button switches, as shown. Call them A, B and C.

You show by pushing switch A that it turns the bulb on and off. Punching the other two switches has no effect on the light. Challenge someone to locate the switch that operates the bulb. Naturally he pushes A, but the bulb stays dark. Allow him a second chance. He tries, say, C. Again the light remains off. Only B remains. It operates the light!

Like a three-shell game of find the pea, this can be repeated over and over. The control switch mysteriously moves from one button to another. Invariably the victim fails on his first two tries. Always the last switch he selects turns the bulb on!

Very few physicists who see this for the first time can guess what seems to be a complicated circuit pattern under the board. It couldn't be simpler. The switches and the bulb are joined in series to a battery.

When all switches are on, the bulb lights. Select any button, say B. Punch it several times to show that it controls the bulb. Leave the switch off; then show that switches A and C have no effect on the light by pushing each an *odd* number of times. This leaves all the switches off. The victim's first choice of a button fails to light the bulb, but leaves that switch on. His second failure also leaves the switch on. Now his third choice is sure to close the circuit and light the bulb.

Repeat the swindle by showing that the last switch he selected does indeed operate the bulb. Punch it several times, leaving it off. Show that the other two switches do not light the bulb by punching each an *odd* number of times. Once again, the victim locates the control switch only on his third attempt.

Try this on a physicist or even on an electrician. After a half-hour of utter perplexity, she will be amazed when you let her see the board's underside.

☞ The Electric Pickle

Cut off the fixture at the free end of an extension cord, expose the two wires, and wrap each around the top part below the head of a large nail. Push the nails into the opposite ends of a dill pickle, the largest you can obtain, and plug the cord into an electrical outlet. Believe it or not, one end of the pickle will glow like an electric light! The glow has a yellow color and is strong enough to be seen in daylight and to read by in darkness.

The glow is believed to come from sodium in the salt used to cure the pickle, but why the glow is always at one end of the pickle is still not clear.

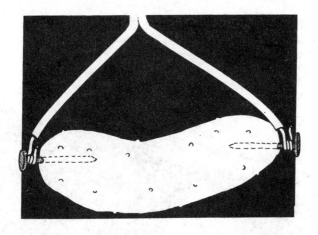

The Human Battery

O ne of the most striking of all stunts with static electricity is that of making a fluorescent light glow without connecting it to a current. Rub such a bulb briskly on your clothing and see it light up from the static charge that the friction produces! This experiment is particularly effective if you perform it at night in a darkened room.

Sound

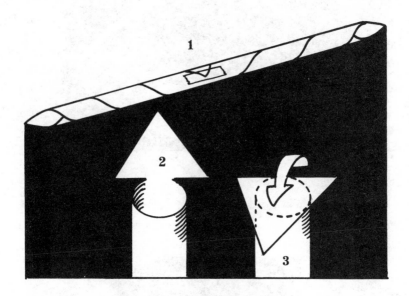

A Puzzling Moo Horn

Here's a simple way to make a reed instrument from a square of paper. Place a pencil at one corner of the sheet, and roll the paper along a diagonal to form the tube shown in 1. A piece of cellophane tape on the outer corner keeps the tube from unrolling. Shake out the pencil.

At either end, snip the paper to form the little triangle shown in 2. Fold the triangle over the opening, as in 3. If you put the other end of the tube in your mouth and inhale, the triangle becomes a vibrating reed that produces the sound of a mooing cow. When others try to blow the horn, not knowing you inhaled, it won't work.

Before handling the "horn" to someone to try, cut a few inches off the mouth end for sanitary reasons. After it fails to work, take it back and "blow" it again by drawing in air. Because the tube is now shorter, the tone rises in pitch. Cupping your hands over the reed, then opening and closing them, makes the sound even more moolike.

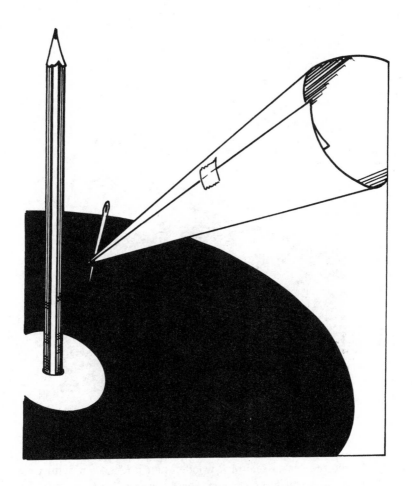

Music from Paper

Push a pencil stub through the hole of a phonograph record. You can spin it like a top on its eraser end.

Fold a square of stiff paper along a diagonal, open it out and crease it to make a sharp point at one end of the creased diagonal. Spin the record and use the paper point as a "needle." The music will be faint but audible.

Louder music can be heard by rolling a sheet of paper into a cone, using tape to keep it from unrolling. Push a needle through the pointed end as shown in the figure. The needle produces a better sound than the pointed paper, and the music is amplified by the cone.

Mysterious Spirit Raps

G rasp the eraser end of a wooden pencil as shown, with your forefinger pushing down on the eraser and your thumb pressed firmly against the pencil's side. Rest the point of the pencil against the top of the table. Push the pencil firmly against the table, and press your thumb as hard as you can against the pencil's side. By moving your thumb upward, you can produce a series of loud raps.

Tap the pencil three times against the table; then produce three raps as if "spirits" are replying. The "spirits" can answer questions from spectators: two raps for yes, three for no. Instead of a table, you can produce similar raps by pressing the pencil's point against the side of a wall. If your thumb is too moist, a bit of rosin on it will help. Friction causes your thumb to move in little jerks that cause the raps.

A Talking Machine

T wist a flat rubber band twice around a wooden spool as shown, and adjust it so it completely covers one end of the spool's hole with the other end of the hole uncovered.

When you blow through the uncovered end, the spool makes a sound like a horn. Now cup your hands around the covered end of the spool. By opening and closing your hands while blowing through the uncovered end, you can make the spool say "Ma ma!"

Why it works: The rubber band vibrates to produce a sound just as the vocal cords in your larynx vibrate when you force air past them. Opening and closing your hands changes the quality of the sound to produce "ma ma" in the same way that you produce a "ma ma" sound by opening and closing your mouth when you speak!

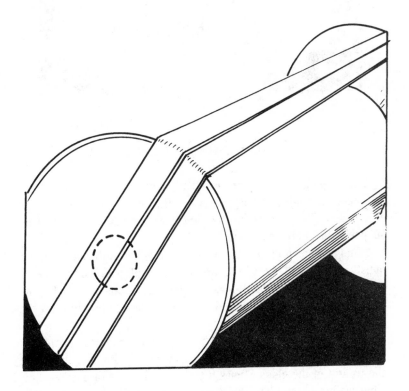

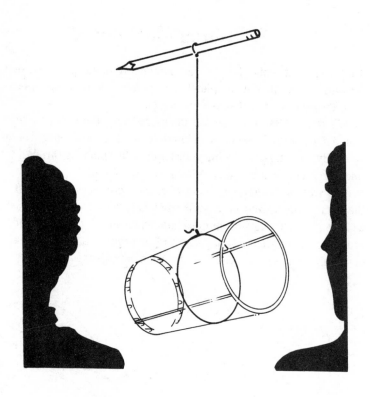

The Ghostly Glass

T he magician ties a piece of string around an empty drinking glass. The other end of the string is then tied tightly to the center of a pencil. She then holds the pencil by one end, with the glass suspended from it, as shown.

"This is a spirit glass," the magician tells her audience. "Ask it a question and it will give you the answer by rapping once for yes, twice for no." Someone asks a question; then everyone listens for the answer. A faint rapping sound is heard, coming directly from the glass!

How is it done? The magician simply turns the pencil very slowly in her hand. At first, the string will start to turn with the pencil; then the weight of the glass causes it to slip back to its former position. This produces a sudden vibration of the string, which is amplified by the glass to produce the mysterious "rap."

Light

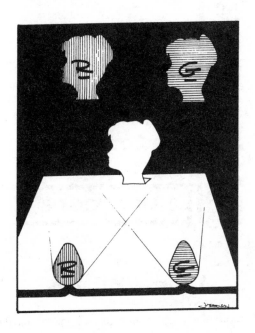

Colored Shadows

Place a red and a green bulb (Christmas-tree bulbs work fine) so they cast two shadows of an object on a wall or screen, as shown in the drawing.

Cut a profile of a face out of cardboard and hold it as shown to project two shadows on the wall. Before doing this, ask someone to guess what colors the shadows will be. An unthinking person will guess that the green bulb will cast a green shadow and the red bulb a red shadow. Of course, it is just the reverse.

Now ask what will happen to the shadows if you turn the face so its nose points toward the light sources. Not many will guess that one shadow will turn until the two faces are looking away from each other. And if you turn the cardboard face until its nose points toward the wall? The shadows will look toward each other!

A Mirror Paradox

Using red and blue crayons, pencils, or pens, prepare two sheets of paper as shown. On one sheet print in capitals WHAT A COOKIE. The words should run horizontally with COOKIE in red and the other two words in blue. On a second sheet, the same words are printed vertically, this time with COOKIE in blue and the other two words in red.

"Why is it," you lie to your audience, "that words printed in blue are always reversed by a mirror, whereas red printing is never changed?"

Hold the first (horizontal) sheet up to a mirror, but as you do this, surreptitiously turn the sheet upside down. Only the red COOKIE appears unreversed. Now hold the other sheet (with the vertical writing) up to the mirror. Again, the two red words are unchanged, but COOKIE, in blue, is reversed!

How come? Your explanation is a good way to introduce people to the concept of mirror-reflection symmetry. Thanks to Marvin Miller for this one.

The Pulfrich Illusion

Tie a small weight to one end of a 2-foot (60 cm) length of string. Stand at one end of a room and swing the weight back and forth like a pendulum.

A person stands on the other side of the room so that the plane of the pendulum's swing is perpendicular to her line of vision. She holds one lens of a pair of dark glasses over one eye, but keeps *both eyes open*. The weight will seem to travel in an ellipse. If she shifts the dark glass to the other eye, still keeping both eyes open, the weight appears to go around the ellipse the other way!

This startling illusion is named for Carl Pulfrich, a German physicist who described it in 1922. He correctly surmised that the dimmer image on one retina reaches the brain a trifle later than the brighter image on the other retina, so that one eye sees a time-retarded image of the moving weight. When the brain combines the two images, it thinks it sees the weight shifted toward or away from you, depending on the direction the weight is moving.

Every now and then a firm advertises special spectacles that it claims will produce 3D images when you watch TV. The glasses are merely a cheap pair of sunglasses with only one dark lens. Horizontal motions across the TV screen seem to be either in front or back of the screen. In Japan, children are given Pulfrich spectacles for viewing animated cartoons that are carefully plotted so there is a lot of horizontal motion to produce an illusion of depth.

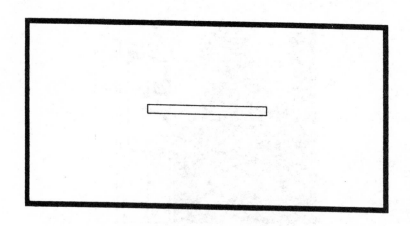

Retinal Retention

A motion-picture film is a series of stills that flash in rapid sequence with split-second pauses between each picture. Television screens also display a sequence of still pictures scanned rapidly. Our eyes do not detect the pauses between the stills.

The illusion of smooth, continuous motion depends on the fact that our retinas retain images for many milliseconds after the image source is gone. A neat way to demonstrate this is to cut a narrow slot in a sheet of cardboard as shown. Hold it over a picture such as the cover of a magazine. You see only a tiny part of the picture. Now move the cardboard rapidly up and down. Thanks to image retention on your retinas, the entire picture comes clearly into view.

A Blacklight Code

Here's a simple way to produce writing that can be seen only under ultraviolet (black) light. Many laundry detergents advertise a power to make clothes bright. They contain a chemical that is activated by the sun's ultraviolet rays. Take a bit of such detergent and mix it with water. Too much water will weaken the effect; too little makes a paste that is visible on the page.

With a clean pen point or a tiny brush, write your message on dark paper or cardboard (white file cards work well). Avoid white paper that also shines under black light. Although invisible in ordinary light, your "secret" writing will glow milky white under ultraviolet radiation.

Shortwave ultraviolet light can damage eyes and skin, so use a bulb that emits long-wave radiation.

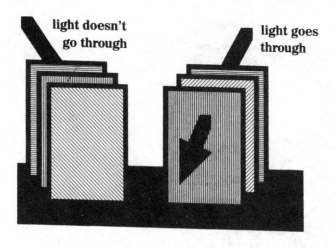

light doesn't go through

light goes through

A QM Paradox

It is well known that if you put together a filter sheet that polarizes light vertically and one that polarizes horizontally, light will not go through. If you put a sheet that polarizes diagonally, at a 45-degree angle, on either side of the vertical and horizontal sheets, light still will not go through.

Now put the diagonal sheet *between* the other two. Light comes through! Don't ask me why. Ask someone who knows his QM (quantum mechanics).

Another Mirror Paradox

Stand in front of a mirror and with a piece of soap outline the image of your face. What will be the height (or width) of the oval you draw, compared with the height of your actual face?

The surprising answer is that the image is half the height of your face. Contrary to intuition, this ratio is constant regardless of your distance from the mirror. If you stand 30 feet (9.1 m) away and outline your face's image with a piece of soap on the end of a 30-foot (9.1 m) pole, the oval will still be exactly half the size of your face! A pleasant exercise is to prove this constant with plane geometry.

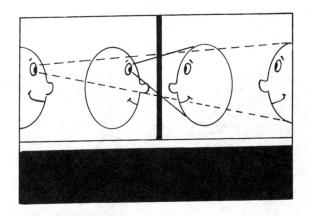

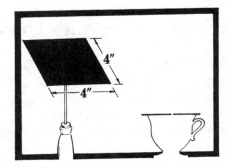

Blacker than Black

C an a white surface appear blacker than a black sheet of paper?
Yes, and this simple experiment proves it.

Find the blackest sheet of paper you can and cut out a square about 4 inches (10 cm) to a side. With an ice pick or sharp pencil point, make a tiny hole in the center of the square. Place the paper on top of a porcelain cup that is all white inside. Under a lamp, the hole clearly is blacker than the paper!

No black paper or cloth is totally black—only a very dark gray. The cup covered by the paper is almost a "black body"—an object that absorbs all radiation. Light entering the tiny hole can exit only after undergoing so many reflections inside the cup that almost no radiation can escape.

69

Sensory Illusions

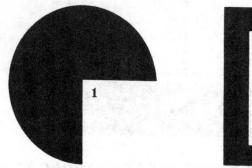

1

2

A Square That Ain't There

To prepare this amusing "betcha" you need a sheet of black paper. Fold it in half twice, then draw a circle about 3 inches (7.5 cm) across on the folded paper. If you don't have a compass, trace around the rim of a circular object. Cut through all four thicknesses of paper to produce four black disks. From each disk, cut out a quarter section, as shown in Figure 1.

Place the mutilated disks on a white surface in the manner shown in Figure 2. A white tabletop is ideal, but a large sheet of white paper will do as well. Because your brain always makes the *best bet* in interpreting perceptions in the light of past experience, you will see what looks like a white paper square on top of the disks. Call someone into the room and bet that he or she can't pick up the square before you count to ten. The illusion is so strong that the square ghost even seems a different shade from the white surface beneath it.

Where Does the Water Go?

F or this trick you need two wine glasses with the tulip shape shown in the picture. Before showing the trick, fill one glass to the brim with water; then carefully pour from it into the other glass until each holds the same amount. It is hard to believe, but both glasses appear almost full.

Then call someone over and bet that you can pour all the water from one glass into the other without any overflow. It looks impossible, but of course you can do it easily.

You might follow this with a joke bet that you can make a match burn under water. Light the match, pick up one of the glasses of water, and hold the burning match under it.

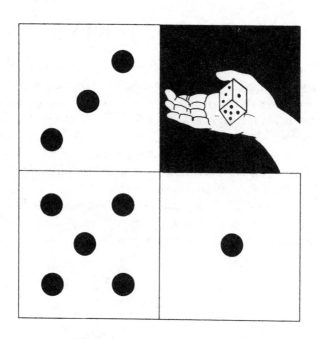

The Enchanted Die

With black ink, draw the pattern shown on thin cardboard. Each die face should be about 3 inches (7.5 cm) on the side. Cut out the pattern, fold it, and tape the edges to make the faces of an *inside-out* die (the dots are inside). Place it on a table, stand across the room, and close one eye. Your brain is so accustomed to seeing cubes that the die will soon assume a normal shape. Keeping the eye closed, move a few feet to either side. The die seems to rotate in a direction opposite to your motion!

Hold the model on the palm of a hand while you stare with one eye at its far corner. In a moment, your brain will snap the die from concave to convex. Now turn your hand in different directions. The die seems to float above your hand, always twisting in directions opposite to the direction your hand turns.

I was told that a psychologist painted the three sides of a die, each about a foot square, on the ceiling and two walls that met at a top corner of his office. Visitors could stand across the room, close one eye, and see the die rotate as they moved from side to side.

Left- or Right-Eyed?

Here's a simple test of whether one of your eyes dominates the other.

Hold two identical pencils vertically in your left hand. Look between them, focusing on something across the room. You will see four images of pencils. Move your hand forward and back until the two middle pencil images merge to form a single pencil between the other two.

Keeping your vision focused beyond the pencils, tap each of the three images with your right forefinger. If your right eye is dominant, you will tap pencils A and B, but when you try to tap C, the finger will go through the pencil as if it were a ghost. If your left eye dominates, image A becomes the ghost.

Another way to demonstrate eye dominance is to put two pennies on the table, about ½ inch (1 cm) apart. As you look down, focus past and through the pennies. Each coin is doubled and you are likely to see four images of coins. Move your head up and down until two of the four images merge and you see only three coins. Feel each image with your finger. You will feel two of them, but the third penny is a ghost.

Still another way to test eye dominance is to hand a person a megaphone and ask him or her to hold the wide side up to his eyes and look directly at you through the small end. The eye *you* see through the small end is the dominant eye!

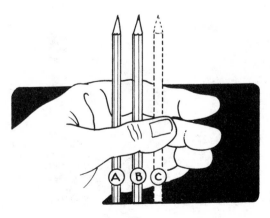

The Bent Playing Card

If you stare at a waterfall for a minute or so and then look away, everything momentarily seems to be rising. We've all had the similar experience of looking out the window of a moving train or car. When the vehicle stops, the scenery for a few seconds seems to drift the other way. Psychologists once believed this illusion was caused by the tired eye muscles, but this conjecture was shot down by a spiral illusion. Stare for a while at a rotating spiral, then look at someone's face. The face seems to expand or diminish depending on which way the spiral turned. Obviously something in the brain, not the eyes, causes this illusion.

Can such an illusion be produced without motion? The surprising answer is yes. Obtain two identical playing cards. Stand one against something on the table so that you view it from an angle, as shown. Hold the other card between your thumb at bottom, fingers at top, and give it a strong bend so you see the face as concave. View this curved card from the same three-quarter angle as you see the unbent card.

With the aid of an egg timer, stare at the bent card for a full minute, letting your eyes roam around its center; then shift your gaze to the unbent card. For a few moments, it will appear slightly *convex*! Something in your brain seems to tire and reverse the bend. No one knows how or why.

Zombie Glass

Bartenders serve "zombie" cocktails in tall, narrow, cylindrical glasses like the one shown on the left in the figure. If you can't obtain a zombie glass, look for a tall glass with proportions like the two in the picture. The zombie glass I have before me is 7 inches (18 cm) tall and 2½ inches (6.4 cm) in diameter at the brim.

Bet someone that the distance around the rim exceeds the height of the glass. It is impossible to believe, but true. This is easily demonstrated with a piece of string or a paper strip you can wrap around the glass's rim.

The circumference at the top of your glass is, of course, pi (3.14) times its diameter. Pi times 2.5 is a little more than 7.8, so the rim exceeds the glass's height by more than ¾ inch!

An Illusion of Weight

Weigh a large empty can. Find a tiny can and fill it with sand until its weight exactly equals the large empty one.

Ask someone to pick up each can separately, then tell you which is the heavier. It is impossible not to believe that the small can is much heavier than the large one. Indeed, most poeple refuse to believe the equality until you prove it by using a weight scale.

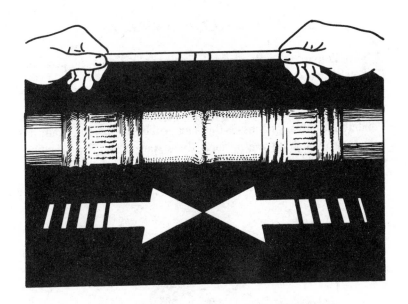

Magnetized Pencils?

Most everyone remembers the childhood stunt of standing between the sides of an open doorway and pressing the back of each hand against the frames. Keep up the pressure for a minute or two, then step out of the doorway. Your hands automatically rise in the air like wings.

Here is a less well known demonstration of the same muscular effect. Hold two pencils horizontally, as shown, and press their eraser ends together. Keep pressing with as much force as you can for a minute or so. Now separate the pencils a few inches. They will tend to move together exactly as if they are bar magnets attracting each other.

This reminds me of a puzzle. Suppose you are given two steel bars, exactly alike, except one is a magnet and the other is not. How can you tell which is which? Of course you are not allowed to touch either bar to any metal object. I won't spoil your "aha!" insight by giving the answer.

The Puzzling Snap

Blindfold a friend, then snap your fingers near his right ear and ask him to tell you which direction the sound came from. He will guess correctly. Snap your fingers near his left ear. Again he will guess correctly. Now snap your fingers near the back of his neck, or beneath his chin. His guesses this time will be wildly off the mark! If others are watching you perform this experiment, you can amuse them by calling out "Correct!" each time your friend makes a bad guess.

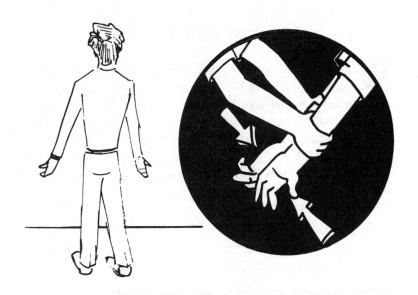

Touching Hands

Have someone stand with his back to you and his arms extended to the sides as shown by the figure at the left in the picture. Tell him to keep his arms straight while you seize each wrist and pull his arms slowly backward. Keep pulling until the backs of his hands touch, just as the boy in the picture is doing. For some strange reason, when his hands come together, it will seem to him that they are still a foot or so (30 cm) apart! You may find it hard to convince him that the backs of his hands really touched. He'll think you deceived him by pressing the back of each hand against something else. The only time he'll really be convinced is when he tries it on someone else and sees what actually happens.

Funny Brush-Off

W ant to give someone the brush? Ask a friend to turn his back to you while you brush his back with a whisk broom. After you do this once or twice, instead of brushing his back, run the whisk broom down the front of your own clothing. At the same time, move your left hand downward over his back, as shown in the above picture. It will feel and sound to him exactly the same as it did before, when you brushed his back. You will find that he is completely unable to tell you when you have brushed his back with the broom and when you brushed it with your hand.

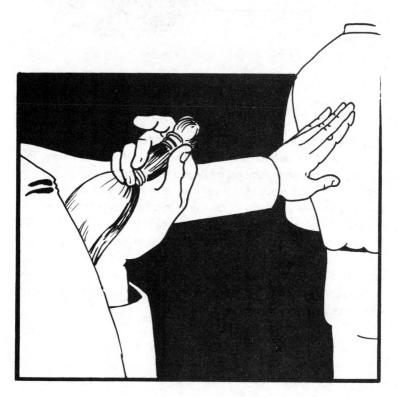

Sloping Teeth

Take a pencil and put it beneath your tongue, as shown in the picture. Bend your tongue downward over the pencil and feel the inside of your lower teeth with the tip of your tongue. It will feel as if your teeth are slanting in the direction of your throat at a sharp angle!

Multiplying Marbles

Tell a friend to cross his middle and index fingers as shown on the left and to close his eyes. Hold out your hand with two marbles on the palm. Ask him to feel the marbles with the tips of his crossed fingers, and tell you how many marbles there are (right). He'll probably guess "three" or maybe "four," and will be astonished when he opens his eyes and sees only two. Even a single marble feels like two when the fingers are crossed!

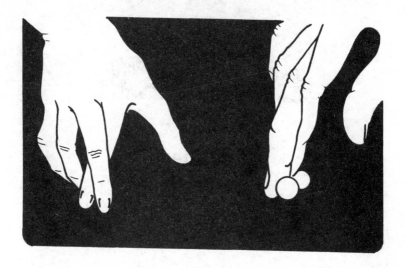

Probability

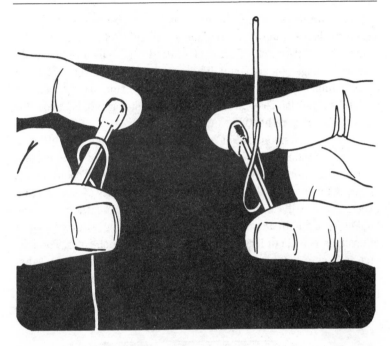

Whirling Wire

S traighten a paper clip except for one end. Press the wire in at that end to make a closed loop. Push a paper match through the loop and hold the match's ends between a thumb and forefinger. Turn the match as shown at left so its wide sides are parallel with the floor, but do not call attention to this fact.

With a middle finger, flick the wire to make it spin rapidly around the match. When it comes to rest, it will hang downward as illustrated (left).

So far, nothing unusual has occurred. Now hand the match and its hanging wire to someone, but when she grasps the ends of the match, see that the match is turned so that its wide sides are *vertical*. Bet that if she flicks the wire with her middle finger, the spinning wire will come to rest pointing *upward*.

Although the odds seem heavily against it, you can't lose! (Thanks to Tom Rodgers of Atlanta for this one.)

The Bunch Effect

B eans flung over a table never distribute themselves evenly. They form large clumps. Statisticians call this "clumping" or "the bunch effect." If you shuffle a deck of cards, the reds and blacks do not mix homogeneously. You will be surprised by the sizes of red and black clumps. The heads and tails of a flipped coin, or the colors of roulette spins, also come in clusters. It is not surprising to encounter runs of six or more heads or tails.

A beautiful demonstration of clumping is obtained by filling a beaker with thousands of little beads, half of one color and half of a contrasting color. From candy stores you can obtain tiny colored balls of candy that come in different colors. Buy a quantity of, say, red and green. Mix the little spheres thoroughly, then pour them into the beaker or a small glass. Look through the sides; you will see a marvelous mosaic of irregular large clumps of each color. The pattern is such a glaring departure from what you would expect that many physicists, seeing it for the first time, suspect some sort of electrostatic effect that binds like-colored beads together.

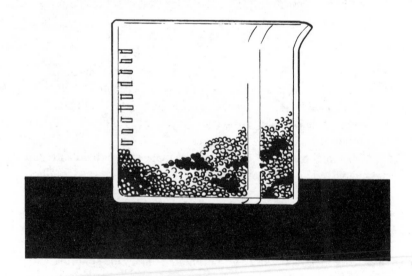

A Probability Swindle

Here is a simple game with three playing cards that seems fair to a player. Actually, the odds strongly favor you, the game's operator.

To play the game, remove from a deck an ace and two cards that are not aces. Shuffle the three cards and place them face down in a row on the table. The victim puts her finger on a card she guesses to be the ace. The probability that it is the ace is, of course, ⅓. You now peek at the other two cards, without letting her see their faces, then turn face up a card *not* an ace.

Two face-down cards remain. One of them surely is the ace, so the probability her finger is on the ace seems to have risen from ⅓ to ½. You bet even odds (fifty-fifty) that her finger is not on the ace. It seems like a fair bet. Actually, the probability that her finger is on the ace stays at ⅓. Because you can always reverse a non-ace, turning it face up has no effect on the game's odds. If the game is repeated many times, you will win two out of three games.

The strange thing about this bet, very difficult for most people to comprehend, is that if she were to change her mind after you have turned up a non-ace, and switch her finger to the *other* face-down card, the probability that it is the ace rises to 2/3. In other words, if she is allowed to switch her choice after you turn over a card, *she* will win in the long run, two times out of three!

Surprising Dice Bet

A sk someone to name two faces of a die. Suppose he names 2 and 5. Let him throw a pair of dice as often as he wishes. Each time you bet at even odds that either the 2 or the 5 (maybe both, but at least one) will show. Because there are four other faces that could come up, it seems like a foolish bet on your part.

There are 36 different ways that two dice can fall. If you make a list of all 36, you may be surprised to find that 2 and 5 will show in 20 of them. Therefore the probability you will win is 20/36 or 5/9, which means you will win in the long run five times out of every nine tosses. The odds are 5 to 4 in your favor!

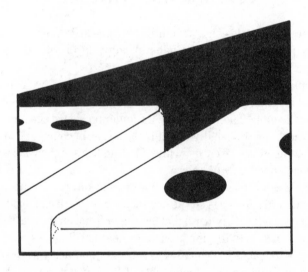

❖❖❖❖❖❖❖❖❖❖❖❖❖❖❖❖❖❖❖❖❖❖❖❖❖❖❖❖❖❖❖❖❖❖

Are You Psychic?

W hile your back is turned, ask someone to jot down 100 *H*s and *T*s, choosing each letter at random as if they were flips of a fair coin that falls either heads or tails. With your back still turned, ask the person to call out the letters in the order she jotted them down.

You will try to guess, by ESP, whether the *next* letter to be called is *H* or *T*. After each guess you are told if you guessed right or wrong.

It would seem that you would be wrong about as often as right. Actually you can score 60 or more hits!

The secret is simple. After each time you guess *right*, call the *other* letter for your next guess. And after each guess you call *wrong*, call the *same* letter the next time. The reason you will make more hits than misses arises from the fact that people tend to ignore long runs of the same letter. In other words, after a short run of letters they tend to switch to the other letter. You would fail to score more hits than misses if the list of *H*s and *T*s were produced by actually flipping a coin.

Dollar Bills

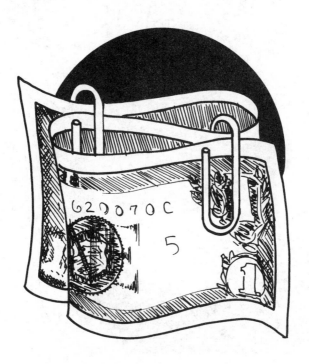

Linking Paper Clips

Attach two paper clips to a folded dollar bill in the manner shown. (Front and back should look identical.) Ask someone to grasp the bill at both ends and pull it straight. Explain that the clips will jump off the bill and travel several feet before they fall to the floor, so the bill must be held so the paper clips point away from the person. Contrary to laws of probability, you continue, the clips will fall so close together that they will actually touch each other.

You are right. The clips link together as they pop forward!

See George Smile and Frown

F old a dollar bill, preferably a crisp new one, to produce a sharp mountain crease that goes vertically through the center of Washington's mouth. Pull the bill flat. The crease will remain as a slight ridge through Washington's face.

Hold the ends of the bill in your two hands, Washington facing you. Tip the bill backward and you'll see George frown. Tip forward and you'll see him smile.

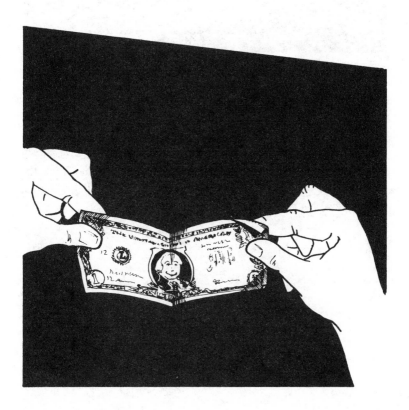

Put George to Sleep

Would you like to see George fall asleep? Make a horizontal mountain fold that goes through Washington's eyes. Crease it sharply between the nails of your thumb and index finger. Unfold the bill. Hold its two ends and tilt it slowly backward. You'll see George slowly close his eyes.

Turn George Upside Down

Start by holding a dollar bill at each end with Washington right-side up. Fold down the top half (A). Fold the right side of the bill *back* (B). Fold the bundle in half again, but this time fold the right side *forward* (C).

Wave a hand over the bill and say "Abracadabra." You now appear to unfold the bill the same way you folded it, but now you cheat a bit. For the first unfold, move the *front* part of the bill to the right (D). Unfold the front to the right again (E). Open the bill by raising the front half (F). Washington will be upside down! Challenge a friend to duplicate the trick. Chances are she'll be unable to do so, even after watching you do it many times.

Support a Glass

B et that you can place each end of a crisp dollar bill on the rims of two side-by-side glasses in such a way that the bill will support a third glass on its middle.

The secret: Make five or six sharp lengthwise pleats in the bill. It will then be rigid enough to support the third glass, as shown.

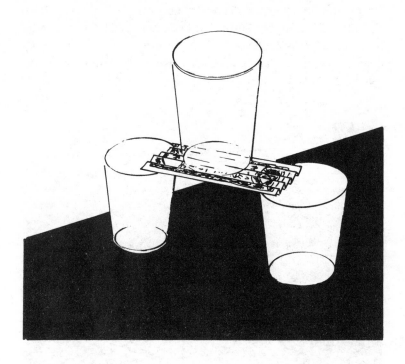

The Balanced Half-Dollar

C rease a bill in half lengthwise, then fold it like a V. Place a half-dollar on it as shown at left. Believe it or not, if you pull slowly and gently on the ends of the bill to straighten it out, the half-dollar will remain balanced on the creased edge!

How Many Eyes?

C hallenge a friend to tell you how many eyes are on a dollar bill. Most people count two eyes on Washington, the eye at the top of the pyramid on the bill's back, and (if they are observant) the eye on the eagle. "No," you say, "the answer is *eight.*"

To find the other four eyes, fold the ends of the bill as shown to make two bug-eyed green monsters.

Blow It Over

Fold a crisp bill in half, then bend down two diagonally opposite corners to make it stand on a table like a tiny platform. Challenge your friends to blow it over on its back. After they give up, you show how. Blow the bill across the table until it projects halfway over an edge. Now get down below the tabletop and blow up on the bill to turn it over.

Index

ace, probability of finding, 85
air and air pressure, 11, 13, 18–22
balancing, 31, 32, 34, 48
balloon: inflated with open end, 20–21; pin trick with, 22
bear, climbing, 49
Bernoulli's principle, 19, 20
blacker than black paper, 69
black light code, 67
book: and hanging threads, 43; balancing, 48
brushing, sensation of, 80
bunch effect, 84
candle, balanced, 23
cards: adhered by air pressure, 18; falling, 29; illusion of bending, 74
carton with water holes, 17
center of gravity, 31, 32
centrifugal force, 14, 43
coin: falls into bottle, 39; guessing heads or tails, 86–87; moving, 38; picked up by straw, 10; supported by dollar bill, 93
compass needle, turning, 53
convection, 27, 28
cork: sinking, 13; lands on end, 33
cups, swinging, transfer energy, 46
dice, combinations of, 86
die, concave-convex, 72
dime: dropping in a bottle, 39; picked up with straw, 10
displacement of water, 9, 11, 16
dollar bill tricks, 88–95
egg, turns when stopped, 40
eggshell: floating, 9; spins on plate, 37; stabbing with knife, 36
electricity, 55–57
energy transfer, 46
eye: dominant, 73; on a dollar bill, 94
face, image of, in mirror, 68
fire, 23–25
feedback, 50
fluorescent, lit by static electricity, 57
friction, 48–54
gas, propels match, 24
glass: amount of water in, 71; supported by dollar bill, 92; zombie glass, 75
gravity, 29–33
hands, touching, 79
heads or tails, guessing, 86–87
heat, 26–28
hoop, bottle, and dime, 39
horn: paper, 58; spool, 61
keys balance matchbook, 34
knots close simultaneously, 50
light, 63–69
linking paper clips, 88

magnetism, 51–52
marbles: illusion of many, 82; rise to top of mixture in glass, 41; spinning, 15
matches: Cartesian, 7; extinguished behind glass, 19; lands on side, 33; launched from match folder, 24; lights from wrong end, 25
matchbook: and keys, 34; roly-poly, 35
matchbox, dropped, 33
mirror paradox, 64, 68
motion: and inertia, 34–47, illusion of, 66
olive, rotating, and centrifugal force, 43
paper clip: magnet removes from bottle, 54; suspended in air, 51
paper cup, floating, 52
paper curls, from moisture, 14
paper money tricks, 88–95
paper "motor," turned by body heat currents, 27
paper phonograph needle, 59
papers, tearing apart, 45
pencils, "magnetized," 77; rapping, 60
penny: falls on heads more often, 30; choosing, by body heat, 26
pickle, electric, 56
polarized light, 68
pool balls and nail trick, 42
probability, 83–87
Pulfrich illusion of depth, 65
rapping pencil, 60
ruler, tipped by air current, 20
sensory illusions, 70–82
shadows, colored, 63
silverware, balanced, 31
soap, as surfactant, 8
sound, 58–62, 78
"spirits": moving compass, 53; rapping with pencil, 60; in rotating glass, 62
spool horn, 61
square, illusion of, 70
static electricity, 57
stick, finding center by feedback, 50
surface tension of water, 10
switches, electrical, 55–56
teeth, illusion of slanting, 81
toys, displace water, 16
type, in mirror, 64
volume, by displaced water, 16
water, 7–17; displaced, 9, 11,16; surface tension, 10; water level, and glass of marbles, 12; pressure, and speed of stream, 17; in wineglass, 71
weight, illusion of, 76
wire, whirling, 83
zombie glass, 75